Nicolas Brasch

Why Do Spiders Live in Webs?

and other questions about habitats

WHY IS IT SO?
?
Science

CAMBRIDGE
UNIVERSITY PRESS

Why do spiders live in webs?

How do otters sleep?

Contents

Questions about habitats

Q: What is a habitat?

A: A habitat is the place where an animal or a plant lives. It lives here because this place has everything the animal or plant needs to survive. It has the right food, water, shelter, sunlight and minerals for it.

Q: What happens when a habitat changes?

A: When a habitat changes a lot, it stops giving an animal or a plant everything it needs to survive. The animals or plants have to move or die. Changes to habitat can be caused by natural reasons, for example, a **drought** or a **flood**. Or they can be caused by human action, for example, people cutting down trees or making **dams** in rivers.

a dam

More questions
about habitats

Q: why do spiders live in webs?

A: Spiders make webs to catch food. They eat flying insects, so they make their webs in places where insects like to fly. Not all spiders live in their web. Some spiders make more than one web and then they come back to their webs to eat their **prey**.

Q: Where do otters sleep?

A: Otters sleep on their backs on the water. When the water is moving, they have to stop themselves from floating away. They usually wrap themselves in seaweed, or sometimes two or more otters hold paws! Baby otters sleep on their mothers' stomachs.

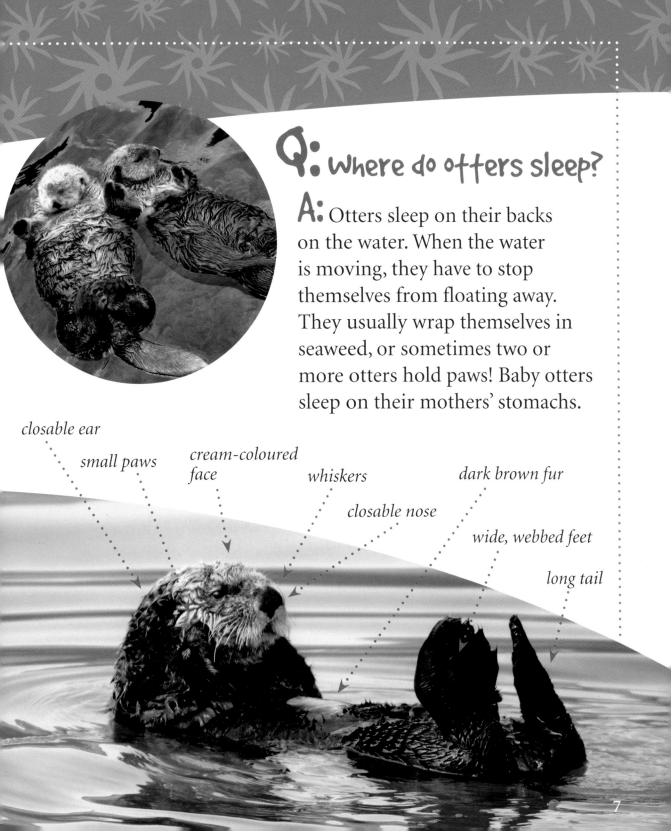

closable ear

small paws

cream-coloured face

whiskers

closable nose

dark brown fur

wide, webbed feet

long tail

> Koalas

Koalas live in a very specific habitat. They only eat eucalyptus leaves. There are 500 types of eucalyptus leaves, but koalas only eat six types!

> Bees

Bees live in **beehives**. They use lots of hexagonal cells of beeswax to make the hives. The cells are called 'honeycomb' and bees use them as homes for young bees that are still growing, and as stores for pollen from flowers, and for honey to eat.

Arctic Ocean

Atlantic Ocean

Pacific Ocean

> The most common habitat

The most common habitat for creatures is the sea. The Arctic, Atlantic, Indian, Pacific and Southern Oceans form about 70 per cent of the Earth.

> Rabbit warrens

Rabbits live in warrens. These are tunnels which rabbits dig in the ground. There can be 100 entrances to a warren in one square kilometre.

> Stampeding animals

When animals **stampede**, they do not destroy habitats. They make them larger. Stampeding disturbs plant seeds. The plant seeds then move to new parts of the habitat.

Tropical rainforests

Tropical **rainforests** are a habitat for many different creatures which live in different parts of the trees or on the ground. Rainforests are being destroyed very quickly. In one second, rainforest the size of two football fields is destroyed.

9

Can you believe it?

a camel

Hibernating bears

Some bears **hibernate** because they cannot find food in the winter.

Before the winter, the bears eat lots of food. When winter arrives, they find a cave or a hole in a tree and go to sleep. Their heart rate slows down while they are asleep and the fat from the food they ate keeps them alive. During hibernation, they are alert enough to respond to danger.

a hibernating brown bear

camel humps

Camel humps store fat from food. When camels are in the desert for a long time and there is no food or water, they use the fat in their humps. Camels can also save water because they do not sweat much when it is very hot.

Magnetic termites

Magnetic termites are insects that live in the far north of Australia. They make nests in lines. The nests face north and south. This keeps the nests cool and protects them from the sun.

magnetic termite nest

Wildlife website for kids:

http://kids.nationalgeographic.com/Animals/CreatureFeature

Chimpanzees: Jane Goodall

The English zoologist, Jane Goodall (1934–), has observed chimpanzees in Tanzania, Africa, for more than 40 years. She discovered that chimpanzees can make and use simple tools. Before that, scientists thought that only humans could do this. Her observations have changed how scientists understand both chimpanzees and humans.

a chimpanzee

Dinosaurs:
Luis and Walter Alvarez

The American father and son team, Luis (1911–1988) and Walter (1940–) Alvarez, made an important discovery that could explain why dinosaurs became **extinct**. They found that the **clay** where the dinosaurs lived contained **iridium**, which is also found in meteorites. They said that a meteorite hit Earth and destroyed the dinosaurs' habitats.

It's quiz time!

1 Match the homes to the animals.

1. nest

2. hive

3. web

4. warren

5. cave

a) bee

b) rabbit

c) termite

d) bear

e) spider

2 Which is the odd one out? Why?

1. drought, trees, flood, fire _____

_____

2. camels, spiders, dam, otter _____

_____

3. water, shelter, rainforest, food _____

_____

3 Choose the correct words.

1. Camel humps contain (water / fat / vegetables) that help them survive long periods in the desert without food or water.

2. Some bears hibernate because they are unable to find food in their habitat during (winter / autumn / summer).

3. Otters sleep on their (sides / stomachs / backs) on the water.

4 Complete these sentences with the words.

> animal   chimpanzees   hexagonal   koalas
>
> plant   eucalyptus   habitat

1. _____ only eat certain types of _____ leaves.

2. A habitat is a place where an _____ or a _____ lives.

3. Jane Goodall studied _____ in their natural _____ .

4. Beehives are made up of _____ cells.

Glossary

beehive: container where bees live and where honey is collected

clay: sticky, dense soil

dam: barriers that stop water flowing so that the water can be collected

drought: a long period without rain

extinct: destroyed as a species

flood: a lot of water in an area that is usually dry

hibernate: to spend winter resting or in an inactive state

iridium: a hard, silvery-white metal

prey: victim

rainforest: forest in a tropical area where it rains a lot

stampede: a herd or pack of fleeing animals